Application of Blockchain Technology
Sector Review of Blockchain Solutions

Application of Blockchain Technology
Sector Review of Blockchain Solutions

Isa Nsereko

Copyright © 2018 by Isa Nsereko
All rights reserved. No part of this book may be reproduced, scanned, or distributed in any printed or electronic form without permission.
First Edition: February 2018
ISBN: 978-1-98-021318-5

To My Wife and Kids:
You Are the World to Me

Table of Contents

Acknowledgement 2

Preface 3

The Blockchain 5

Financial Services 10

Education 14

Healthcare 18

Insurance 22

Real Estate 26

Energy Sector 29

Further Applications of Blockchain 32

Conclusion 35

References 36

9. https://www.velox.re 36

10. https://www.lantmateriet.se 36

11. https://powerledger.io/#intro 36

12. https://www.brooklyn.energy/about 36

Acknowledgement

I would like to express my gratitude to the many people who saw me through this book; to all those who provided support, talked things over, read, offered comments, allowed me to quote their remarks and assisted in the editing, proofreading and design.

I would like to thank Kindle Direct Publishing (KDP) for enabling me to publish this book. Without you this book would never have been possible.

Above all I want to thank my wife, and the rest of my family, who supported and encouraged me as I wrote this book in spite of all the time it took me away from them. It was a long and difficult journey for them.

Last and not least, I beg forgiveness of all those who have supported me in all ways over the course of the years and whose names I have failed to mention.

Preface

Since the turn of this century, a number of technological developments have taken place that have affected the way we are destined to do things for the next few decades.

From Apple Inc.'s iPod, to Skype, to YouTube, and to Virtualization technology, one such technology that is set to cause even more massive disruption is the Blockchain Technology or simply referred to as blockchain.

Blockchain technology gained prominence after the launch of the cryptocurrency, bitcoin. Since then, blockchain has been at the forefront of creating cutting edge solutions across a number of sectors. From the way the technology has been used, it appears that there are endless possibilities in the application of blockchain.

It also goes without saying that the growth of blockchain technology has come with numerous challenges with regard to regulation, risks and safety concerns.

This book provides readers with insights into how different sectors are exploiting or leveraging the blockchain technology to not only address key challenges they face today but also how blockchain can be applied in the future.

This book is not meant to be exhaustive in reviewing the possibilities that blockchain solutions provide.

Rather it is meant to provide readers with insights into current and future developments of blockchain technology.

This book also aims to get readers to think about how the blockchain can impact their lives, businesses and their sectors.

Organization and reading of this book:

The book is organized into eight chapters. A brief description of the chapters follows:

Chapter 1 puts blockchain technology into context for readers and discusses its importance and risks thereby setting the scene for its implementation across various sectors.

Chapters 2 to 7 take an in-depth analysis of how sectors ranging from financial services, education, insurance, healthcare, real estate and the energy sector are exploiting and leveraging blockchain solutions to address key challenges in those sectors.

Chapter 8 looks at further application of blockchain solutions across more sectors.

This book makes for easy reading. Readers are encouraged to read the introduction first as it lays the foundation for the rest of the book.

After the introduction, readers may choose to read any of the remaining chapters 2-7 in any order of their interest or preference of sector.

Lastly the chapter 8 provides a wider perspective of blockchain solutions driving innovation in more sectors.

Happy Reading!

Isa Nsereko

(B.Sc., MBA)

Chapter 1

The Blockchain

INTRODUCTION

Hardly a day goes by without breaking news on the internet and major news channels on blockchain technology and its related developments. News stories are awash on new break through developments as well as fortunes made or fortunes lost in the rapidly changing blockchain technology environment.

After several months of personal research, driven by my inquisitive mind's desire to follow new technologies, I would like to share this knowledge with readers who are keen to keep abreast of developments in blockchain technology sphere that is taking the world by storm and how these developments are likely to affect us.

My research has given me deep understanding of blockchain technology and its potential application across various sectors of the society. In my own opinion It is would be difficult to explain blockchain technology without understanding the underlying technology that supports the blockchain, known as Distributed Ledger Technology (DLT).

The term Distributed Ledger Technology (DLT) refers to a group of technologies where a ledger is maintained by a group of peers (hence distributed) rather than a single central authority like a stock exchange, central bank, national payments system or even a government. In its simplest terms, a ledger refers to a place where data transactions or entries are recorded and stored.

The blockchain hence is one type of distributed ledger which enables records to be sorted and stored into blocks. The growing list of records is referred to as a block. Blocks are hence a collection of records or data. Every time a transaction takes place, it is added to other transactions onto a block. Whenever one block of data is filled up with transactions, it is stored on the blockchain and a new block is created hence forming a chain of blocks!

The blockchain is maintained by a network of communicating computers called nodes and each node will independently verify and store its own record of each and every transaction.

In short, Distributed Leger Technologies are a generic name for the family of technologies underpinned by a distributed ledger. The blockchain is just one possible application of Distributed Ledger Technologies.

The history of blockchain may be traced back before the beginning of this millennium. However, blockchain became popular from the end of 2008 when Satoshi Nakamoto published the white paper for bitcoin.

Blockchain is most known for being the technology behind cryptocurrencies and other platforms.

Software developers soon realized that the underlying blockchain technology could be separated from cryptocurrencies and used for other innovations across a multitude of sectors.

WHY BLOCKCHAIN TECHNOLOGY?

Reliance on a single central authority has its own merits but has also raised a number of concerns including data integrity, security, speed of processing transactions, intermediation costs as well as privacy concerns.

The risks associated with a single central authority have to some extent been mitigated using backups and restricted security access to databases.

Blockchain technology has been fronted for providing the following benefits to its users:

IRREVERSIBLE AND IMMUTABLE

No transaction record can be deleted from the blockchain and any invalid transactions are easily identified and rejected. All transactions are encrypted and trust is enabled through consensus protocols allowing trusted exchange of value.

QUICKER TRANSACTION PROCESSING

The doors to financial and payment institutions are only open for business during normal business hours. Coupled with the challenge of operating across various time zones, the end result is a delay in transaction processing times.

Blockchain technology is operational 24, 7 without the constraints of business operating hours nor differing time zones. Blockchain transactions are hence processed and settled faster.

TRANSPARENCY AND ACCOUNTABILITY

Blockchain technology uses open source protocol which means that other developers and users have the opportunity to modify the protocol as they deem fit and necessary based on consensus.

The key benefit that comes with open source protocol is that it is a particularly secure technology because it makes any alteration of records within the network very difficult as everyone on the network can see what has been altered.

DECENTRALIZATION

The blockchain is not managed by a single central authority or organization, hence it is decentralized. This prevents a single point of failure in its operations.

The other benefit decentralization offers to the blockchain is that it allows individual transactions to have their own proof of validity and the authorization on the network, thereby improving data integrity and minimizing the risks of data being compromised. Any deviations from the original rules of protocol are immediately recognized.

USER-CONTROLLED NETWORKS

The blockchain relies on its users, investors and developers to reach majority consensus on deciding the outcome of any future developments on the network and any other controversial issues.

LOWER TRANSACTION COSTS

Blockchain technology permits peer-to-peer and business-to-business transactions on the network to be completed without the need for a trusted third party.

What this means is that blockchain has the potential to actually reduce costs to users on the network.

RISKS ASSOCIATED WITH BLOCKCHAIN TECHNOLOGY

Blockchain technology also poses certain risks that include the following:

ABSENCE OF REGULATION

A number of governments are still in early stages of reviewing and putting in place legal frameworks and regulations that will govern blockchain technology.

NEW TECHNOLOGY

Blockchain is still a new technology and is yet to stand the test of time in its application in a number of sectors. Whereas the technology has already provided ground breaking solutions, it is still early to make judgment on its continued application.

KNOW-YOUR-CUSTOMER (KYC) REQUIREMENTS

Until framework and regulations are put in place, establishing origin of funds on the blockchain will be challenging. Also monitoring usage of funds will pose a challenge for authorities.

HIGH VOLATILITY

The high volatility recently observed in some of the cryptocurrencies has resulted in gains for some investors but also huge losses as well. Investors in cryptocurrencies need to be aware of this.

SECURITY CONCERNS

The recent spate of thefts, scams, breaches and hacks at some cryptocurrency exchanges and on some highly rated Initial Coin Offerings (ICOs) has raised a number of concerns for participants and investors in blockchain driven technologies.

THE FUTURE OF BLOCKCHAIN TECHNOLOGY

Being a relatively new technology, it may be too early to judge the success or otherwise of blockchain innovations.

The recent stories of the dark web related to blockchain has raised debate over whether the failures are due to the technology itself or due to misuse of this technology.

Despite of the risks that come with blockchain technology, a number of sectors are already experimenting with a number of blockchain driven solutions individually or in cooperation with other organizations and institutions.

We shall next look at how blockchain technology is being used to address key challenges across different sectors.

Chapter 2

Financial Services

The global financial services industry is still recovering from the financial crisis of 2008. The industry is dealing with increased regulation, more sound risk management practices, focus on talent, better cost management strategies and increasing innovation to sustain profitability and deliver on all other stakeholder expectations.

In pursuit of further growth, the financial services industry is devising means of reaching out to its customers in a cost effective and more efficient manner. The financial services industry relies on technology for much of its innovation and blockchain technology is already being explored and applied within the financial services industry.

Blockchain Technology offers a number of benefits to the financial services industry including enabling greater access to financial services, greater transparency, lower costs, and faster transactions.

SIMPLIFY KNOW YOUR CUSTOMER (KYC) PROCEDURES

Know your Customer (KYC) and Customer Due Diligence (CDD) regulations are intended to help reduce money laundering and terrorism activities. Financial institutions are required to verify and identify their clients. Blockchain allows for independent verification of a client by one institution and then this verification can be accessed by other institutions on the network.

This way the KYC process wouldn't have to be repeated each time a client requires to access services from another financial institution on the blockchain network. The reduction in administrative costs for compliance departments would be significant.

FRAUD REDUCTION

Financial Institutions and their intermediaries are targets of economic crime. Most banking systems are built on a centralized database. The centralized database offers a single point of failure that can be vulnerable to cyber-attacks.

As a distributed ledger, the blockchain records and locks transactions thereby providing transparency and accountability as users can access the full historical data easily.

Blockchain has the potential to tackle financial crime. This technology would eliminate some of the current crimes being perpetuated online today against financial institutions.

TRADING PLATFORMS

Trading platforms relying on blockchain-based technology would reduce the risk of operational errors and fraud, while improving processing efficiencies.

PAYMENTS

Financial institutions use a number of third party providers in the processing of making payments between themselves and their clients. Blockchain could eliminate these intermediaries, making the payments process more secure and less costly.

SMART CONTRACTS

Smart contracts are essentially computer codes or programs that are self-executing provided certain conditions are met. Blockchain can store computer codes that are automatically executed once two or more parties enter their keys.

Smart contracts can be useful in the area of Trade Finance. Using blockchain technology, financiers, trading houses, and any other trusted intermediaries are able to see when the goods have been shipped and can release funding appropriately.

Blockchain could be used in documentary business. This reduces time taken to confirm assets, confirm transactions, release payment and make confirmations.

CASE USES OF BLOCKCHAIN:

[1]*R3*

"R3 is an enterprise software firm working with over 100 banks, financial institutions, regulators, trade associations, professional services firms and technology companies to develop Corda, our distributed ledger platform designed specifically for financial services".

"Launched in September 2015, R3 was born out of a common frustration amongst banks and other financial institutions with multiple generations of disparate legacy financial technology platforms that struggle to interoperate, causing inefficiencies, risk and spiraling costs".

"Recognizing the power of distributed ledger technology lies in its network effect, we worked with the industry to build the largest collaborative group of its kind in financial markets".

[2]*UBS AG*

"One of the great promises of blockchain technology has been in speeding up and streamlining the trade and settlement process, increasing efficiency and potentially massively reducing costs.
In August 2016, UBS announced that it was teaming up with Deutsche Bank, Santander and BNY Mellon, as well as the fintech company Clearmatics, to promote UBS's "utility settlement coin" (USC). The USC is a blockchain-based digital currency that financial institutions could use to directly transact securities with each other, bypassing the traditional settlement process. Initially developed by Clearmatics in conjunction with UBS, the USC is intended to be directly convertible into central bank cash. With it, banks could significantly reduce the time and cost of post-trade settlement and clearing."

Financial Institutions are clearly exploiting and leveraging blockchain technology. Financial institutions will need to review a number of issues in embracing this technology including data protection concerns, privacy and confidentiality issues, different regulations and legacy systems.

Chapter 3

Education

The global education industry is experiencing rapid growth due to the realization of its importance in economic development. The Education sector is enjoying increased focus and government support, as well as increasing enrolment numbers over the years.

Technology is also changing the way education is being delivered. A number of new learning methods are being used by both the youth and the adult segments.

Blockchain technology has a role in shaping the future of education. Education still relies on reputation and trust. Due to globalization, learners seek education opportunities across the globe. Employers also need to keep abreast of the changes in education and to position themselves to understand and manage education offered on the global scale.

The blockchain can impact the education sector in the following ways:

ISSUING, STORING, AND AUTHENTICATING CERTIFICATES:

Educational institutions can use blockchain technology to store and deliver their issued academic certificates. Encryption is used to create, sign-off and place certificates onto the blockchain database. Employers can then access and authenticate the issued certificates. This measure would quicken the time taken by employers to verify certificates and also to stop the practice of using fake certificates to acquire employment.

Educational institutions could also use blockchain to create and use shared repositories of certifications. These repositories could be used by affiliated educational institutions or groups of schools, providing them a cheap shared resource for easy use and access.

CREATION OF A NATIONAL DATABASE

Education systems and curricula vary from country to country and from region to region. Educational institutions hence offer credentials at various levels of accomplishment within the system.

The blockchain could offer a standardized and shared approach where the full range of credentials that are being produced at all levels in the system (schools, colleges, universities, institutes, examination boards, trade associations, and employers) can be stored securely and accessed by stakeholders.

Blockchain could offer a centralized but neatly distributed national database for the authentication of both process and certification for vocational education and apprenticeships.

GLOBAL ASSESSMENTS

The current system of certification is largely paper based. This creates the risk of loss, damage and even fraud. Fraud can occur through creation of fake documentation or by alterations.

Current education systems may require students to move from one academic institution to another, from one country to another, from one job to another or to collect education credits from different institutions.

Blockchain could provide a secure, online repository that could be helpful in storing and managing global assessments whilst providing easy authentication of the assessments.

Mass Open Online Courses (MOOCs) are growing in popularity as a learning method. MOOCs issue their own certificates. The blockchain could offer secure certification to the major MOOC providers.

CPD

For professions, continuing professional development (CPD) is key to maintaining membership and credibility. Tracking of CPD is challenging as there is no standard format that can be used and accessed by both professionals, associations and employers.

Blockchain provides a platform for issued CPD data from conference attendance, courses, and other forms of learning by capturing those experiences and learning opportunities in a securely stored and reputable system.

Blockchain can be used for storing achievement of training of employees through its more open and secure system. The achievements can be accessed internally by employees and also when they leave an organization.

OPEN BADGES

Open Badges are another flexible and portable way to recognize learning. They are increasingly being used alongside traditional qualifications and professional accreditation.

Blockchain offers a tamper-proof system that can be used for storage of open badges and for dealing with their authentication.

CASE USES OF BLOCKCHAIN:

[3]GRADBASE: INSTANTLY VERIFY QUALIFICATIONS

"LET'S END CV FRAUD. FOR GOOD.
1/3 of people lie on their CV. Worst, embellished and fake CV can make honest candidates miss out on their careers opportunities.

So, we introduced a new global standard for issuing and instantly verifying qualifications. Powered by the Bitcoin Blockchain."

[4]*BLOCKCERTS*:

"Blockcerts is an open standard for creating, issuing, viewing, and verifying blockchain-based certificates. These digital records are registered on a blockchain, cryptographically signed, tamper-proof, and shareable. The goal is to enable a wave of innovation that gives individuals the capacity to possess and share their own official records".

Blockchain technology clearly has applications in the world of learning at all levels. As blockchain continues to revolutionize education, players will have to review data-regulation concerns as well as deal with culture that has been centered around individual education institutions hence the need to change current thinking.

Chapter 4

Healthcare

The global healthcare industry is projected to register stable growth in the coming years with high growth opportunities within emerging markets. Stakeholders are looking for innovative and cost effective ways to provide access and quality to health care.

Healthcare is another industry experiencing the blockchain revolution. Stakeholders are already reviewing the possibilities that blockchain can bring into the healthcare industry now and into the future.

Global challenges for the healthcare industry include the need to institute a common database of health information that doctors and providers could access irrespective of their electronic medical system, possession of legacy medical systems, complex billing processes, the need for higher security and privacy of patient records, overpriced medical tests, useless treatments, and medical fraud.

Other concerns in healthcare include the time spent by doctors on administration compared to that spent on patient care. There is also a need for better sharing of research results to facilitate new drug and treatment therapies for disease.

Blockchain can be used in the following ways in the healthcare industry:

DATA SECURITY

With the growth of connected devices and the Internet of Things, existing health IT architecture is struggling to keep systems secure. Blockchain solutions can keep health data private and secure whilst reaping the benefits of connected medical devices.

MEDICAL DATA MANAGEMENT

Blockchain technology can be used to store electronic medical records and at the same time allow for its secure access to other medical providers who need to use the records. This saves precious time, removes unnecessary costs and prevents duplication of procedures and resources.

Blockchain can be used for the improvement and authentication of health records. Blockchain could reduce barriers involved in complex data sharing agreements between hospitals, physician providers, and public health departments whilst making sure that data gathered is reliable and current across all parties.

DRUG DEVELOPMENT

Considerable amounts of data are produced during clinical trials. This includes safety and quality reports, statistics, blood tests, surveys, medical imagery and so on. There could be mistakes committed whilst handling this data.

Blockchain technology can give patients and medical providers a one-stop access to their entire medical history across all providers they have ever visited. Using blockchain technology and with the permission of patients, researchers can gain access to medical records for faster medical innovation.

SUPPLY CHAIN INTEGRITY

Blockchain technology could help reduce the multi-billion counterfeit drugs market by tracking all transactions between drug-makers, wholesalers, pharmacists and patients to verify and secure drug product information.

CLAIMS AND BILLING MANAGEMENT

False claims and poor billing causes billions of losses globally. Such losses can be minimized by using blockchain technology.

Blockchain can provide automation and more efficient processing in the system. Blockchain can also reduce administration related billing costs by eliminating the need for intermediaries.

CASE USES OF BLOCKCHAIN:

[5]*iSolve*

"iSolve is dedicated to helping the BioPharma and Healthcare community succeed through collaboration, partnerships, and technology. Our solutions create opportunities for companies to advance their drug development and patient care initiatives through innovation and growth. Our company has 2 very distinct lines of solutions for supporting these industries with the Blockchain. The first is to enhance and innovate the Drug Development Lifecycle and the Drug Supply Chain by providing secure data sharing and visibility across the entire Drug Supply Chain from API to Patient in order to improve patient outcomes. The second is to support the acquisition of IP Assets (SMART IP) and provide a marketplace for providing services for initiatives secured by the Blockchain. In the long term, we see ADLT™ as the key to redefining the movement and use of data throughout the healthcare ecosystem."

[6]*THE MEDILEDGER PROJECT*

"The MediLedger Project is a collaboration between Chronicled and The LinkLab, bringing together expertise in both Pharmaceutical Supply Chain and Blockchain technologies. Our intention is to advance the dialogue of a blockchain utility to enhance Pharma companies' ability to manage their supply chains."

"We believe a Pharma industry blockchain utility and ecosystem can:
- Improve compliance of track and trace regulations
- Improve patient safety and drug supply security
- Enable easier reconciliation of exceptions
- Provide a platform for further business transformation for processes associated with the transfer of ownership of prescription medicines"

The Healthcare industry is rapidly embracing blockchain technology which is expected to provide better support for their innovations and operations. Other future innovations and applications are under development.

Chapter 5

Insurance

The global insurance market was also affected by the financial and economic in 2008. Currently, the industry is experiencing modest growth of insurance premiums as it recovers and as the global economy returns to normal growth.

The key challenges faced by the insurance industry include managing insurance fraud, slow processing times for claims, regulatory issues and low insurance penetration in certain markets.

Blockchain can help resolve some of these insurance industry challenges. The blockchain will certainly change the way the insurance industry operates by creating more trust and efficiency. The insurance industry relies on trust management.
The blockchain can impact the insurance industry in the following ways:

IMPROVE THE CLAIMS PROCESS

Blockchain technology can improve the claims process through use of smart contracts. The smart contracts contain rules, regulations and mechanisms for actions that need to be taken in the event that certain conditions occur.

The smart contracts do not require any human intervention to execute. This would greatly minimize manual claims processes, and speed up the process. There would also be cost savings from reduced administrative work.

ENHANCE EFFICIENCIES

The data entry process for taking out insurance can be cumbersome and so is the process of switching insurers. Keeping control over your personal data is an issue of concern for customers.

Blockchain technology provides a solution to the above challenges by providing security that would allow the personal data to be controlled by the client while verification is registered on the blockchain.

SMART CONTRACTS

By nature of the legal issues to be covered, insurance contracts are usually lengthy with a lot of details in the fine print. This is done by the insurer to ensure they are adequately covered against circumstances and to mitigate against fraud.

The blockchain provides a solution through the use of smart contracts whereby both the insured and insurer benefit from managing claims in a responsive and transparent way.

The insurance contract would be recorded and verified on the blockchain. Once a claim is submitted, the blockchain ensures that only valid claims are paid. In the event of multiple claims being logged for the same event, the network would identify and reject such claims.

FRAUD DETECTION AND PREVENTION

Fraudulent insurance claims increase insurance premiums within the market. Blockchain technology has the potential to detect and prevent fraudulent activity.

The blockchain technology's decentralized repository allows for validation. Historical records on the blockchain can be used to independently verify customers, policies and transactions for authenticity. This requires extensive cooperation between insurers, customers and any other parties.

The blockchain can be used to share information to prove existence of contracts, purchases of products, and to verify police reports for insurance purposes.

IMPROVE TRUST

Lack of trust, high costs and inefficiencies all contribute to the low levels of insurance penetration. The blockchain can build trust in insurance because it provides transparency of records and transactions for both the insurer and the insured.

CASE USES OF BLOCKCHAIN:

[7]*B3i:*
"INSURERS AND REINSURERS LAUNCH BLOCKCHAIN INITIATIVE

Swiss Re together with a group of Europe's biggest insurers have joined forces in an attempt to put Blockchain through its paces to evaluate whether the technology can help make the insurance industry more efficient.

Aegon, Allianz, Munich Re, Swiss Re and Zurich have launched the Blockchain Insurance Industry Initiative B3i aiming to explore the potential of distributed ledger technologies to better serve clients through faster, more convenient and secure services. If Blockchain technology proves viable, it could well streamline paper work and reconciliations for reinsurance and insurance contracts and accelerate information and money flows, while greatly improving auditability."

[8]*Smart Contract-Based Insurance Policy:*

"NEW YORK - 15 Jun 2017: American International Group Inc. (NYSE: AIG), IBM (NYSE: IBM) and Standard Chartered Bank plc announced they have successfully piloted the first multinational, "smart contract" based insurance policy using blockchain, a distributed ledger technology."

"The blockchain solution creates a new level of trust and transparency in the underwriting process, enabling AIG and Standard Chartered to deliver multinational insurance more efficiently. Coordinating management and placement of multiple insurance policies across multiple countries is highly complex. The pilot solution was built by IBM and is based on Hyperledger Fabric – a blockchain framework and one of the Hyperledger projects hosted by The Linux Foundation."

Blockchain technology has the potential to solve a number of key challenges in the insurance industry. Blockchain can minimize human intervention through smart contracts. This will improve speed, efficiency, and security within insurance and also lead to a shift in the way people buy insurance in the future.

Chapter 6

Real Estate

The Real estate industry has gone through a tumultuous period in recent times affected by macroeconomic trends such as interest rates, population growth, and economic strength.

Real estate is a cyclical industry. The industry grew tremendously after the World War II economic boom and then sank during the inflationary period of the 1970s. The early 1980s so a rise in the real estate industry until the depression at the end of that decade. The industry recovered again by the mid-2000s when residential real estate boomed until the mortgage crisis hit and prices collapsed.

Some of the challenges experienced by the real estate industry include the reliance on a number of paper based transactions, a tighter regulatory environment, bureaucracy, lack of transparency and fraud.

The real estate industry is slowly recovering from its history. Blockchain technology can make real estate activities to become easier, quicker and cheaper. The market would get rid of unnecessary intermediaries, have lower transaction costs, become safer, more transparent and consequently more liquid.

The blockchain is already causing disruption in the real estate industry in the following ways:

MONEY TRANSFERS

The blockchain can be used to make money transfer payments for real estate by transferring conventional (fiat) money issued by central banks and also by cryptocurrencies.

PROPERTY, TRANSACTIONS AND TITLE LEDGERS

The blockchain can be used to record and store information on real estate including transactions, title registration, property encumbrances and their physical condition. The information can be entered into distributed ledgers that are accessible online and through mobile apps.

Each property would have its own unique identification plus all its characteristics specified on the blockchain. This would create a real estate property repository and would also make property appraisal easier and faster.

SMART CONTRACTS

Escrow accounts are often used in buying and leasing real estate. Rental deposits or sale proceeds may be held in such accounts until all conditions pertaining to right of ownership, lease or rental agreements are fulfilled.

The blockchain has the potential to execute property and rental transactions using smart contracts. Buyers can place money in a blockchain escrow account. Smart contracts automatically check the transaction possibility and legitimacy and reject any agreements from being concluded that do not meet pre-established standards.

VOTING

The blockchain can also be used to execute real estate decisions that require voting. Owners of shared infrastructure can vote on decisions such as major repairs or works on common areas. The blockchain would guarantee reliable remote voting and give owners the certainty that their votes have been registered correctly.

CASE USES OF BLOCKCHAIN:

[9]velox.RE:

"velox.RE is the most accomplished blockchain real estate team, building on the most robust blockchain, Bitcoin. The velox.RE - Cook County (Chicago) pilot program produced the first ever legal blockchain deed software and procedural protocol. Blockchain will allow truly digitized property ownership, exchange, and data. velox.RE is the vertical platform that enables blockchain applications, giving real estate stakeholders more transparency, liquidity, and profitability and enable the next evolution of the real estate industry."

[10]Lantmäteriet:

Sweden's land registry authority (Lantmäteriet) has been experimenting with the recording of property transactions on the blockchain. The expected benefits are eliminating paperwork, reducing fraud, and speeding up transactions.

Using blockchain technology, the real estate industry can speed up transactions by reducing the need for paper-based record keeping. It can also help with the secure storing of records, ensuring accuracy of documents, tracking transactions, verifying ownership, and transferring property deeds. The blockchain can create a real estate market where transactions are more secure, transparent, faster and equitable.

Chapter 7

Energy Sector

Global energy needs are likely to grow steadily over the next decade. The demand for energy is driven by economic growth and rising population in developing economies.

The key concerns for the energy sector include accessibility, affordability, and search for renewable and sustainable energy sources, energy efficiency, environmental impact and climate change.

One of the Sustainable Development Goals set by the United Nations is about ensuring access to affordable, reliable, sustainable and modern energy for all. As nations pursue this goal and look for ways to address the rising demand for clean and affordable energy, blockchain technology can play a role in driving change within the energy sector in the following ways:

SECURE STORAGE OF TRANSACTIONS

The blockchain can be used for providing secure storage of ownership records in a register for better asset management in the energy sector. The possibility of storing all transaction data in a tamper-proof and decentralized blockchain can also be used by nations to manage emissions trading.

SMART DEVICES

Smart devices are able to communicate with each other and with other devices inside and outside homes. The transmitting and storing of the related information and transactions can be done on the blockchain.

The blockchain could be used to share data from smart meters in a transparent way and reduce cases of overcharging of energy usage by service providers or to control energy consumption.

The transactions would be initiated and recorded in a tamper-proof way on the blockchain. The blockchain has the capacity to balance energy demand and supply. When there is an excess of energy supply, this is stored and or transferred to where this energy is required.

DECENTRALIZED ENERGY SUPPLY

Today's electricity supply system comprises of many players including producers, transmitters, distributors and wholesalers. The common set up of the electricity sector comprises of massive, centralized power plants that generate power which is then supplied over transmission and distribution lines. There are also a number of private power producers who supply to the main grid.

The blockchain can provide a decentralized energy supply system in which producers are directly linked to consumers without the need for intermediaries. This would reduce energy costs for consumers.

BILLING AND PAYMENTS

The blockchain can provide a billing system where automatic billing is done for energy consumed. The cumbersome and costly process of meter reading could be eliminated while at the same time boosting the accuracy of bills.

Customers can use cryptocurrencies to pay for the energy consumed. Smart meters would measure the amount of energy consumed, and payments will be automatically executed by smart contracts through the blockchain.

CASE USES OF BLOCKCHAIN:

[11]PowerLedger

"P2P TRADING

This class of Platform Application gives retailers the ability to empower consumers (or in an unregulated environment, the consumers themselves) to simply trade electricity with one another and receive payment in real-time from an automated and trustless reconciliation and settlement system. There are many other immediate benefits such as being able to select a clean energy source, trade with neighbors, receive more money for excess power, benefit from transparency of all your trades on a blockchain and very low-cost settlement costs all leading to lower power bills and improved returns for investments in distributed renewables."

[12]*Brooklyn Microgrid*

"Welcome to Brooklyn Microgrid
Brooklyn Microgrid is developing a community-powered microgrid here in Brooklyn. This means our participants can engage in a sustainable energy network and choose their preferred energy sources, locally".

"A microgrid can operate independently from the larger grid during power outages; providing the neighborhood with a safe, resilient energy grid that can serve as a backup in case of emergencies."

Blockchain technology in the energy sector certainly shows a lot of potential and needs to be further developed by market participants. The benefits for all stakeholders are numerous. Existing regulation will require review to open up the space further for more innovations in this sector.

Chapter 8

Further Applications of Blockchain

Blockchain Technology has the potential to disrupt almost any industry. The following are potential areas that could benefit from blockchain technology:

In Supply Chain Management, Blockchain can be used to monitor costs, labor, and even waste and emissions at every point of the supply chain. Blockchain can also be used to verify the authenticity or fair trade status of products by tracking them from their origin. Transactions are documented in a permanent decentralized record, and monitored securely and transparently. This reduces delays and human error.

Blockchain technology can create a decentralized network of Internet of Things (IoT) devices. Operating like a public ledger for a large number of devices, the blockchain can eliminate the need for a central location to handle communications between the smart devices. Devices including smart homes, smart meters, and smart cars would be able to communicate to each other directly to update software, manage bugs, and monitor energy usage.

Crowd funding is a popular method for fund raising for start-ups and projects. The rationale behind crowd funding platforms is to create trust between project creators and project investors. Using the blockchain, trust is created through smart contracts and online reputation systems, which remove the need for intermediaries. New projects can raise funds by releasing their own tokens that represent value and can later be exchanged for products, services, or cash.

Blockchain can be used to create a decentralized peer-to-peer ride sharing app that allows car owners and users to arrange terms and conditions without the need for third-party providers. The use of built-in e-wallets can allow car owners to automatically pay for parking, highway tolls, and electricity top-ups for their electricity powered vehicles.

Data on a centralized server is vulnerable to hacking, data loss, or manipulation. The blockchain can be used for Cloud Storage. Blockchain technology makes cloud storage more secure and robust against attacks.

Cybersecurity can also benefit from the secure platforms provided by the blockchain. Although the blockchain ledger is public, the data is verified and encrypted using advanced cryptography. This way the data stored on the blockchain is less prone to being hacked or changed without authorization.

The administration of charity is another area in which the blockchain can be used to reduce inefficiency and corruption. Blockchain can be used to track donations to ensure they end up in the right hands.

The electoral process is another area in which the blockchain can be used for voter registration and identity verification. The blockchain ensures that only legitimate votes are counted, and no votes are changed or removed. The blockchain would create transparency by creating an immutable, publicly-viewable ledger of recorded votes. This would be a massive step towards making elections more fair and democratic.

Government systems are often described as being slow, opaque, and prone to corruption. Governments can implement blockchain-based systems that can significantly reduce bureaucracy and increase security, efficiency, and transparency of their operations. The blockchain provides a platform for permanent and secure record of transactions thereby enhancing transparency and accountability for government.

The administration of public benefits is another area that suffers from slowness and bureaucracy. The blockchain can help assess, verify, and distribute welfare or unemployment benefits in a much more streamlined, faster and secure way.

The blockchain can securely provide data storage and verification data for telecommunication subscribers. Telecommunication companies can also use smart contracts to manage roaming contracts with subscribers and several service providers for more efficient billing. Telecommunication companies can provide wallets for cryptocurrencies for their subscribers.

The blockchain can be used in research, consulting, analysis and forecasting. Blockchain can be used to place and monitor bets on anything from sports to stocks in a decentralized way.

The music industry has suffered from piracy and copyright concerns over the years. Also the payments for royalties to songwriters and musicians is a rather complex process. The blockchain through smart contracts allows for musicians to get paid directly from their fans, without giving up large percentages of sales to platforms or record companies. The Smart contracts can also be used to automatically solve licensing issues, and to better catalog songs to their respective creators.

In the aviation industry, blockchain can be used for the registration of airplane components. Information on the authenticity of airplane components can be checked on the blockchain.

In the automotive industry the blockchain can be used to track automotive title transfers. The blockchain can also store automotive maintenance records and to track automotive components during the assembly process, reducing the need for automotive mass recalls.

Any industry that deals with data, or transactions of any kind, is a field than can likely be disrupted by blockchain technology. The space is wide open and the opportunities are many.

Chapter 8

Conclusion

Blockchain technology is still in its growth phase and yet to reach widespread adoption globally. The blockchain market is expected to grow in leaps and bounds over the next decade.

Blockchain can offer benefits to users in the form of faster transactions, improved security features, cost-effectiveness, accountability, transparency, immutability and increased access to services.

As more and more sectors embrace blockchain technology, the market will have to deal with challenges such as standardization of records, systems, processes as well as data privacy, regulatory and legal issues for a collaborative approach.

I trust that this book has given you greater insight into the potential of blockchain technology.

I look forward to providing you further in-depth reviews in the near future of more sectors embracing blockchain technology and to keep you abreast of further developments in the sectors already covered here-in.

So long, for now!

References

1. https://www.r3.com/about
2. https://www.ubs.com/magazines/news-for-banks/en/products-and-services/2016/building-the-trust-engine.html
3. https://www.gradba.se/en
4. https://www.blockcerts.org/about.html
5. http://isolve.io/solutions
6. https://www.mediledger.com
7. http://www.swissre.com/reinsurance/insurers_and_reinsurers_launch_blockchain_initiative.html
8. http://www-03.ibm.com/press/us/en/pressrelease/52607.wss-AIG, IBM, Standard Chartered Deliver First Multinational Insurance Policy Powered by Blockchain
9. https://www.velox.re
10. https://www.lantmateriet.se
11. https://powerledger.io/#intro
12. https://www.brooklyn.energy/about

www.ingramcontent.com/pod-product-compliance
Lightning Source LLC
Chambersburg PA
CBHW031556210526
45464CB00003B/1309